TALKING ANIMAL

By David Taylor

TVS

ALAN SUTTON
1985

ALAN SUTTON PUBLISHING LIMITED
BRUNSWICK ROAD · GLOUCESTER

First published 1985

Copyright © David Taylor 1985

ISBN 0-86299-262-1

Printed in Great Britain

Contents

Introduction

This book, like the 'Talking Animal' television series produced by TVS and co-presented by Andrea and me is about magic – the incredible story of animals, how and why they are as they are, their cleverness, their brilliant design. There is no such thing as a dull or boring animal. Every one of the two million or more species alive on earth today is amazingly successful, for they are all survivors, winners in the evolutionary race. Only man by his selfish or unthinking acts can bring the continuing story to a sad end.

David and the Talking Animal cobra.

In the pages that follow, we will try to convey just a bit of the delight and enthusiasm that Andrea and I find in living creatures. I have been a veterinary surgeon now for almost 30 years. As far back as I can remember I always wanted to be a doctor to wild animals. Andrea is an out-and-out animal fanatic with a keen love of horse-riding and Ben, a cheerful black dog, is undoubtedly her best pal.

If, as I suspect, you are a bit of an animal nut too, read on – and let's start Talking Animal!

Andrea and Ranee, the Talking Animal elephant.

1. The Bat

A quarter of all mammalian species are bats; there are 951 different kinds of bat in the world today. The names of some of the species are quite delightful. Among them are Dawn bats, Leaf-nosed bats, Bulldog bats, Thumbless bats, Wrinkle-lipped bats, Fish-eating bats, Red bats, Grey bats, Brown bats, Bamboo bats and Painted bats. Literally a bat for every taste! Bats feed on a wide range of items and each species tends to specialize in one type of diet. Bat menus include insects, scorpions, shrimps, mice, other bats, lizards, frogs, fish, fruit, flowers, pollen and the blood of other animals.

A vampire bat feeding on an unsuspecting donkey.

Some bats are fliers who can migrate over distances as great as 2,000 kilometres. Their wings flap and give them the power of real flight. No other mammal can take to the air in this way. Creatures such as flying squirrels for example, merely glide down from trees helped by expanded flaps of skin along the sides of the body. The bat often uses its wings to catch insects, particularly the bit between the hind legs which it uses as a sort of scoop. Some bats can fly at almost 50 kilometres per hour.

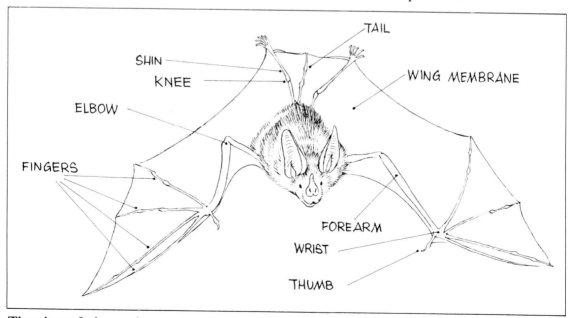

The wings of a bat are broad-winged webs of skin stretched between the bones of greatly elongated forearm, hand and hind leg. The bat flies by means of its enlarged, webbed hands that contain bones, four fingers and a thumb, as a human hand.

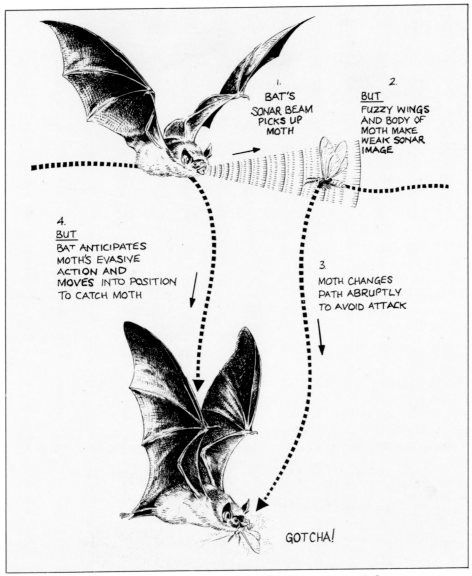

1.
BAT'S SONAR BEAM PICKS UP MOTH

2.
BUT FUZZY WINGS AND BODY OF MOTH MAKE WEAK SONAR IMAGE

4.
BUT BAT ANTICIPATES MOTH'S EVASIVE ACTION AND MOVES INTO POSITION TO CATCH MOTH

3.
MOTH CHANGES PATH ABRUPTLY TO AVOID ATTACK

GOTCHA!

Dog fight: the bat uses sonar to catch a moth which has its own defence mechanisms.

Certain species such as the so-called Flying Foxes possess excellent vision and Fruit-eating bats have a good sense of smell. Most species rely on an incredibly highly-developed hearing ability combined with echo-location. Bats find objects in the dark by echo-location or sonar. They send out a stream of sounds which bounce off objects and are received by their very sensitive ears. The sounds, usually inaudible to the human ear, are emitted from their mouths or, in some species, from their noses. Some bats can continue sending out sounds while feeding, by 'whistling' through a gap in their teeth.

No one knows why bats in deep caves (some caves in Texas can contain up to 20 million of them) do not bump into one another when flying around in the darkness. Somehow they can distinguish their own sound echoes from those of the other 19,999,999 bats!

While some species migrate, many others hibernate during cold weather. They select suitable hide-outs, such as caves or tree-holes, where humidity is high. This is to reduce the need for water during their sleep. Safely installed, they become torpid with a much reduced body temperature and slowed-down breathing and heart rates. This 'go slow' of bodily functions saves energy and conserves fat. Bats go into hibernation with plentiful fat reserves which they have built up during the summer and autumn. When they wake in spring they may have lost one-third of their body weight. During hibernation they wake up from time to time – to urinate! Bats can reach an age of 30 years or more.

One bat, the Vampire bat, lives by drinking the blood of domestic animals and occasionally that of human beings. Vampire bats are very fussy about which victims they attack. They prefer certain breeds of cattle and also select cows rather than bulls and calves rather than adults. The Vampire bat can carry that terrible disease, rabies. Vampire bats inhabit tropical and sub-tropical America. If a British zoo wished to bring some Vampire bats into the country for exhibition, they would have to be kept under *permanent* quarantine conditions.

David takes to the air and hopes to imitate the moth.

The frightening vampire of horror movies is portrayed with many of the features of a real vampire bat.

A pin-up among bats: the head of the Naked bat.

The rarest British bat is Bechsteins bat which is to be found in small numbers in parts of Southern England, particularly the New Forest. All British bats are protected by law, and you must not disturb, harrass or handle them – even if they live in your attic, roof space or cellar. Bats do not get tangled up in your hair when flying at night, nor are they dirty or dangerous animals (the Vampire bat apart). They are diminishing in numbers in the United Kingdom and need our help and sympathy to survive. They are useful animals for the control of insect pests and generally do no harm. In some countries, bats that feed on nectar and pollen are under threat of extinction because of the destruction of their habitat. The absence of their good effect in spreading pollen as they move from plant to plant has lead to diminishing fruit harvests. The West African Iroko tree, the focus of a £60,000,000 a year timber trade, depends on bats for the spreading of its seed. Bat droppings, called guano, form a valuable source of fertilizer in many parts of the world. One American bat cave alone contained 100,000 tons of guano.

Bats are threatened by the destruction of their habitat. The renovation of old buildings, the use of pesticides to treat timber have brought many species to the point of extinction. In some countries, certain tropical bats are killed as a source of high quality meat. Although bats can be found in every part of this planet except the Arctic, they are declining in numbers, and everywhere the culprit, directly or indirectly, is man.

2. The Elephant

In the past there were more than 350 kinds of elephant on earth. Now, only two species, the African and the Asiatic or Indian, remain. The African elephant is found on the African continent below the Sahara. It can stand almost four metres high at the shoulder and may weigh six and occasionally even as much as ten tonnes. The Asiatic elephant is found in India, Malaysia, Indonesia, China and Indo-China. It is not quite as big as its African relative, having a shoulder height of around three metres and a weight of perhaps five tonnes. The Asiatic elephant has the smaller, neater ears while the African sports the bigger triangular ones.

Elephants communicate by using smell, sight, sound and touch. Changes in position, posture and the way in which the head, ears, trunk and tail are held can convey messages, intentions and emotions. Elephant sounds come in all forms from trumpetings and bellows to squeals and growls. They frequently produce a low 'tummy growl' that actually comes from the larynx (voice-box). It can be heard up to two kilometres away and often serves as a warning. Elephants are great touchers and love to explore and caress one another (and human friends) with the sensitive tip of their trunks.

David and Andrea on Ranee, in East Dene, Sussex.

The elephant's trunk is a marvellous multi-purpose instrument that can be used for smelling, touching objects gently, feeding, watering, cooling (by spraying water or sand over the body), breathing, breaking, attacking, greeting, caressing, and trumpeting – a form of communication. The large ear flaps are a cooling device. They act as radiators and can be fanned to increase the cooling effect.

The elephant's foot is broad and the weight is so evenly distributed that it barely leaves an impression upon the ground. The skin is fairly thick (two to four centimetres) but is nothing like as difficult to inject

through from a zoo vet's point of view as, say, a rhinoceros. There is very little hair on the body except at the tip of the tail. The skin is very sensitive – yes, elephants are ticklish! – and elephant skin care in the form of regular mud baths and powdering with dust or sand, is necessary to keep it in good shape. A happy contented elephant purrs just like a cat, though of course the sound is much much louder!

And what about the elephant graveyards – the cemeteries where it is said elephants go to die? I am afraid they are only a legend. Perhaps the finding from time to time of a huge collection of elephant bones can be

Elephants are not afraid of mice! They are short-sighted but have a marvellous hearing and sense of smell. Elephants never run or jump. An adult can eat 200 or 300 kilograms of vegetable food a day, and drink up to 100 litres of water. During the war in Germany, an elephant, probably suffering from a nutritional deficiency, killed and ate a zoo secretary called Bertha Walt.

A rock hyrax, the elephant's nearest relative, wondering what to do next.

attributed to the work of poachers, or the death of a group of elephants in a bush fire or at a water-hole in times of severe drought when no food is available.

The hyrax, the elephant's closest living relative, is really quite a surprise. It is not, as you might suspect, the tapir or the rhino or the hippopotamus; it is a small furry creature a bit like a guinea-pig. Grey or brown in colour, weighing about one to five kilograms and living in Africa in habitat ranging from rocky valleys to high forests, this unlikely distant cousin is actually the Coney mentioned in the Bible (*Proverbs 30 v26*).

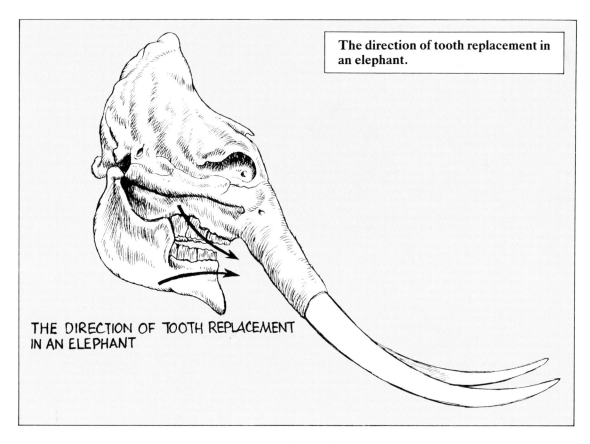

THE DIRECTION OF TOOTH REPLACEMENT IN AN ELEPHANT

The direction of tooth replacement in an elephant.

Elephants have only four teeth in use at any one time. When the two teeth at the front wear down, they fall out and are replaced by two new ones which move forward from the back of the mouth. Throughout its life an elephant can call on a total of 24 teeth. After they are all used up, an elephant in the wild would starve to death. In a zoo or safari park this does not happen, and an old elephant is provided with plenty of soft food such as boiled rice, bran, crushed oats and soft fruit, which does not need chewing.

Both in the wild and in captivity elephants can reach an age of 70 years and perhaps more. The big head of an elephant contains a skull that possesses many air-filled spaces (sinuses) to reduce the weight. An elephant's tusks can grow up to 3.5 metres long and weigh over 100 kilograms each. The ivory of the tusks is a unique combination of calcium, cartilage (gristle) and dentine, a substance that forms the bulk of both human and elephant teeth.

Poachers murder elephants principally to obtain the ivory tusks, although modern plastics are just as good for making trinkets and chess pieces.

14

A baby elephant suckling – an elephant mother's breasts are between her *front* legs.

Elephants possess two breasts or mammary glands for giving milk to their young. These are situated between the *front* legs, not the back as with the cow, goat and many other animals. Elephants have the longest mammalian gestation (pregnancy) period of between 21 and 23 months. Other elephants help in the birth of the baby, acting as 'midwives' by removing the placental membranes and helping the youngster to get on its feet. Most baby elephants are born in the wet season because at that time there is plenty of green food which enables their mothers to produce a large flow of milk. The babies suckle by mouth (not by trunk) for up to four years. A baby is produced every three to four years by an elephant mother and she gives birth to most babies during the period when she is between 25 and 45 years of age.

Elephants live in groups, bull elephants usually living apart from the females and youngsters. A typical family would have two to four sisters with their babies or grandmother, mother and grandchildren. The oldest female is the head of the group.

3. The Cat

There are 37 species of wild cat alive in the world today. They range from the Rusty spotted cat of India and Ceylon through the elegant leopard of Africa to the majestic tiger of Asia. Okay, so you have heard all about the lion and the jaguar and the lynx, but what about the Kod-Kod of Chile and Argentine, the Flat-headed cat of Malaysia and Indonesia or, rarest of all, the Iriomote cat which was only discovered in 1967 on a remote Japanese island?

The domestic cat of Europe and America probably arose from the crossing of two wild species, the Forest wild cat and the African wild cat. The original tame cats were almost certainly tabby, and very similar to the Forest wild cat that still survives in lonely forests in the north of Scotland. The African wild cat is larger and stockier than the modern cat, with a light or orange brown coat and narrow dark stripes. Found in forested regions of Africa and Asia, this animal is said to exist still in parts of the holiday islands of Majorca, Corsica, Sardinia and Crete. The origin of that most superior of cats, the Siamese, is wrapped in mystery.

The stripes on a tiger serve as a most efficient camouflage in the broken light and shade of the forest. No two tigers have identical markings and even the two sides of a tiger are not patterned identically.

The cat has very good hearing, and with 30 muscles working its ears as compared with six in man, it can turn them precisely to locate sounds. The cat is also a great sniffer and very sensitive to certain chemicals. The reason a cat is attracted to plants such as catmint and valerian is that these herbs contain a chemical which is almost identical to a substance released in a cat's urine during the mating season. The cat's nose is also equipped with highly sensitive nerve endings that can detect heat and cold.

The function of a cat's whiskers is not fully understood. It has something to do with touch, and removing them can disturb a cat for some time. Apart from touch, cats are highly sensitive to vibrations. Like some other species, they may give warning of a coming earthquake. Village peasants on the slopes of Mount Etna keep cats as early warning devices of volcanic eruptions.

Purring is not a sort of voice. It is the vibration of blood in a large vein in the chest cavity where the vein passes through a sheet of muscle separating the chest from the abdomen. Contraction of the muscles round the vein 'constricts' the blood flow and sets up vibrations, the sounds of which are magnified by the air-filled bronchial tubes and windpipe. Purring can be heard, but more importantly can be felt. Kittens, born blind, without a sense of smell and with un-developed ears, react strongly to the purring vibrations and make for their mother and the safety and suckling she provides. It is interest-ing to note that feline mothers stop purring as soon as the kittens begin to suckle; the homing device has done its stuff. We do not know why male cats purr.

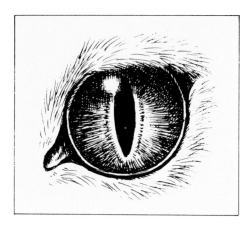

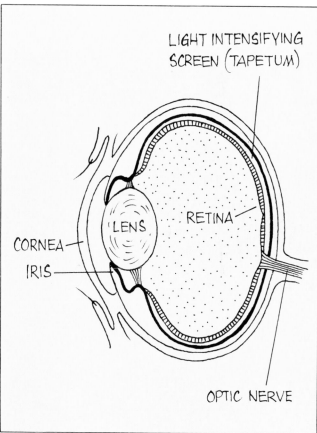

The hunter cat's eyes are perfectly adapted for working in the dimmest light. They possess an ingenious light-intensifying screen or mirror of shimmering crystals set behind the retina. This screen gathers every speck of available light after it has passed through the retina and bounces it back on to the light sensitive layers a second time. This means the retina has the best chance of receiving light. The crystal screen is what makes a cat's eyes flash in the dark. In total darkness, however, cats cannot see any better than human beings. Cats do see colour, but not very well.

David serves lunch to a leopard – cats sometimes relish vegetables.

Although cats are basically carnivorous they are not exclusively so. Among wild cats, the Flat-headed cat is the only species which feeds on vegetation to any extent – it has a particular love of fruit and sweet potatoes. After making a kill, lions and tigers will usually go straight for their victim's stomach, neglecting the prime fillets of meat, to devour the soup of digesting vegetation. The Fishing cat, whose diet consists largely of molluscs, shrimp and fish, is known to have a particular liking for mouthfuls of wild garlic and water weed. Similarly the domestic cat often fancies a nibble of some herb or grass.

The tiger prefers deer, wild pigs and antelopes but will also tackle bears, wolves, lynx, leopards and young elephants. Sometimes it dines on nuts, fruit and berries. All man-eating tigers are disabled, injured or diseased in some way. A typical cause is where a tiger has attacked a porcupine and got quills embedded in its face or paws causing pain and infection and making the hunting of normal prey difficult. It then turns to 'easy' prey in the form of human beings. The domestic cat is as much an instinctive hunter as a tiger or leopard and a hungry cat does not make a better mouser. Indeed, well-fed cats are better mousers. They have the stamina, energy and quick reactions required for the sport, for that is what it is. Mickey, a well-fed tabby of Burscough in Lancashire, killed over 22,000 mice in 23 years before dying in 1968 – an average of nearly three mice a day throughout his life.

The domestic cat can run at speeds of up to 43 kilometres per hour, compared with the 73 kph of a greyhound and the over 100 kph of a cheetah. When walking, cats move the front and back legs on the same side simultaneously. The only other animals that do that are the giraffe and the camel. The household cat does not like to swim. The Persian swimming cat, a rare breed, is said to be more enthusiastic about taking a dip. Among wild cats, the tiger is a strong swimmer as is the Fishing cat from the jungles of Asia. The feline swimming star, however, is undoubtedly the jaguar. What strokes do cats use when they swim? The dog paddle!

Tigers love messing about near water.

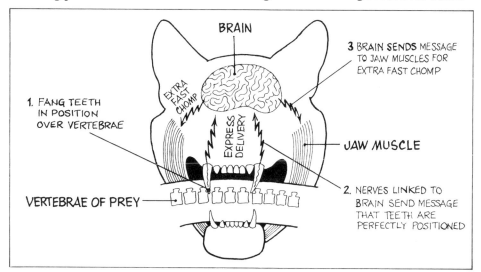

The deadly bite of the cat.

The killing bite of a cat, domestic tabby or jungle tiger, is a remarkably precise affair. All felines tend to use a neck bite. The prey is killed usually by dislocation of the vertebrae in the neck. It is fascinating to note that the distance between the left and right fang teeth of a cat is the same as the distance between the neck joints of its usual prey. A domestic cat has its fang teeth aligned for dislocating the neck of a mouse and the tiger is designed to do the same to its favourite meals, deer and wild pig. There are special nerves linked to the fang teeth of the cat which sense in the twinkling of an eye when the point of the teeth are perfectly positioned over the neck joints of the prey. These nerves then send ultrafast messages to the brain, which responds in turn by sending messages to the jaw muscles instructing them to close at an unusually high speed. The neck bite of a cat is a brilliant computer-controlled process.

Cats have a great homing instinct and often track down their homes despite great difficulties. The longest recorded journey by a cat in search of its human friends is 1,450 kilometres, from Boston to Chicago. Modern research suggests that the key to cat journeying of this sort lies in navigation similar to that employed by birds. It works something like this: during the months or years that the cat was living in its original home, his brain registered the angle of the sun at certain times of day. How does the cat tell what time it is without wearing a watch, you ask? Most animals we believe possess internal biological clocks. Certainly man and the higher mammals have them and they have been located in creatures like cockroaches too.

Suppose the cat is uprooted to a strange place, where the sun's angle at a certain time is slightly different. If he wants to put it right, the cat must use trial and error. In one direction he finds the angle gets worse. He tries another; the angle improves. That must be the direction to go. All this is a subconscious activity, of course, but gradually, in trying to get the sun in the right spot in the sky at the right time, puss finds himself in a neighbourhood where the smells, sights and sounds are familiar. From then on it is plain sailing.

The ancient Egyptians venerated the cat (which they called *miw*) and executed anyone who killed cats, even accidentally. It is no wonder that they were defeated in 500 BC by Persian soldiers carrying cats tied to their shields – no Egyptian would risk injuring one of the sacred creatures. In AD 936, Howel Dda, Prince of Wales, valued cats as follows: 'kittens one penny, mouser two pennies', and a penny was a lot of money in those days! Anyone killing a cat had by law to pay its worth in corn, holding the animal by the tail with its nose touching the ground and then completely covering it with a mound of grain. When Shakespeare's friend, the Earl of Southampton, was imprisoned in the Tower of London, his black and white cat broke into the fortress, sought out the right cell and entered it by shinning down the chimney. A contemporary painting shows the two of them doing 'porridge' together.

Andrea says if you must wear fur make sure it's FAKE.

4. The Dog

Dogs as we know them first came on the scene in Eurasia between 12 and 14,000 years ago. From what kind of animal did they directly spring? The old idea that it was a form of jackal or jackal/wolf cross has been abandoned. Now we believe it was a small middle eastern strain of grey wolf, an animal with a wide variety of coat colours which was distributed throughout Europe Asia and North America at that time.

Man has employed dogs in a rich variety of ways – as guards, hunters, war machines, guide-dogs, rodent controllers, draught animals, footwarmers and provider of hair and meat. Mastiffs in light armour, carrying spikes and cauldrons of flaming sulphur and resin on their backs, were used in warfare by the Romans and in the Middle Ages against mounted knights. Sadly updated, dogs were trained by the Russians in the Second World War to carry out suicide missions against German tanks. They would run between the tracks of the vehicles with mines strapped to their backs. The mine would explode as soon as a vertical antenna attached to it touched the metal of the tank. Carts pulled by dogs were used in Belgium, Holland, Germany and Switzerland until quite recently. Such canine labour was forbidden by law in Great Britain in 1885.

Dogs are still eaten by man in Asia where red coloured specimens are for some reason particularly prized. The Chinese consumed considerable numbers of chows and other red dogs until recently. Australian aborigines

Dogs were often used for hunting in the nineteenth century.

use dingos for warmth on cold nights by sleeping with them clasped in their arms. Aboriginal women, when not carrying young children, often 'wear' a dog draped across the lower back with the head and tail in the crook of their arms as a 'kidney warmer'! The peoples of Mesopotamia used giant mastiffs to hunt lions, while the Jews shunned all contact with what they regarded as an unclean dung-heap frequenting beast. At the present time, around five and a half million dogs are kept as pets in Great Britain. There are over 100 breeds. Although their sizes and shapes vary enormously, all dogs are essentially the same animal, not too far removed from their primitive ancestors. Most of the distinctive characteristics of modern dog breeds have been produced fairly recently by selective breeding by human owners.

The fastest land mammals, the cheetah and certain antelope can only maintain high speed over relatively short distances. Hunting in the animal world is however often carried out over long distances and here is where the persistent long-distance runners of the dog family come into their own. African hunting dogs will pace one another, some loping

The top five breeds of dogs least likely to bite

1. Labrador Retriever
2. Golden Retriever
3. Shetland Sheepdog
4. Old English Sheepdog
5. Yorkshire Terrier.

The runners up are Welsh Terrier, Beagle, Dalmatian and Pointer.

The most popular breeds at the moment

1. Yorkshire Terrier
2. Alsatian (German Shepherd)
3. Labrador Retriever
4. Cocker Spaniel
5. Rough Collie

The Indian Pariah dog, built to survive, with prick ears, medium-sized strong legs and well-developed snout; modern breeds are really artificial creations by man.

behind for a while as other race ahead. When the leaders tire, the lopers move to the front and keep up the relentless pace. After a long chase, this species will actually run down and kill lions. Wolves can achieve speeds of around 56 kilometres per hour.

In the water the dog is no more than an adequate swimmer, but there is one canine who is an excellent swimmer and diver – the Wild racoon dog of China, Japan and Siberia. An expert fisherman, the racoon dog can stay under water for a remarkably long time when in pursuit of his lunch.

Vision is quite well developed in the dog though most species do not hunt primarily by sight and often 'miss' creatures which stand stock still. But shepherds claim that their working dogs will react to purely visual hand signals at distances of up to one mile. Although not totally colour blind, dogs see mainly in black, white and various shades of grey.

Dogs gobble rather than enjoy their food. The canine sense of taste is rather poorly developed.

Dogs are used in France and Italy to find truffle fungus which grows underground, and in Holland and Denmark they are used to detect gas leaks. Everywhere dogs are used to sniff out humans, explosives and drugs, often under incredible conditions. An Alsatian belonging to the Cairo police successfully followed the track of a donkey that had been made four and a half days previously.

Modern dogs evolved from a strain of grey wolf, which is not such a dangerous animal as people think.

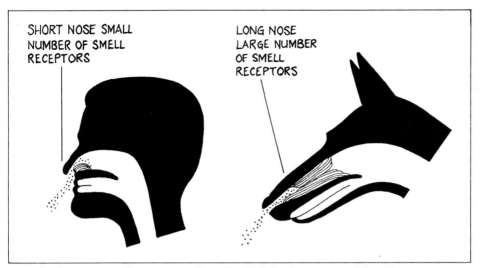

SHORT NOSE SMALL NUMBER OF SMELL RECEPTORS

LONG NOSE LARGE NUMBER OF SMELL RECEPTORS

The dog has a marvellous sense of smell which is about one million times more acute than that of a human being. Only eels are better smellers.

How do they do it? Smells consist of molecules of particular chemicals which float in the air. When these molecules land on the special smelling membrane inside a nose, nerve impulses carry the 'smell information' to a particular part of the brain. This smelling centre is highly developed in the dog and relatively far larger than in man. The smelling area in the human nose is about three square centimetres, whereas in the canine nose it covers almost 130 square centimetres, arranged in anatomical folds which filter the smells from the incoming air. To house such a structure, with some exceptions among 'artificial' breeds, dogs have developed long noses.

Even more important, there are many more nerve cells in the dog's smelling membrane of the nose than there are in a human being's. We have five million smell nerve cells, a dachshund has 125 million, a Fox terrier 147 million and an Alsatian 220 million! A wet nose helps a dog to smell – it dissolves molecules floating in the air, bringing them into contact with the smelling membrane and clearing away old smells.

Every individual be he human, deer or dormouse, has a sweat which is as unique as fingerprints. It is made up of a number of different smelling chemicals. A dog can recognize the 'scent image' of a person and can even make important deductions from the changes of various ingredients of the smell. This allows him to run along a trail for a few metres, register the change in the image and thus determine the direction of travel of his quarry. One of the chemicals present in sweat, butyric acid, is also found in the body of any dead animal and, if eaten, produces an improved sense of smell

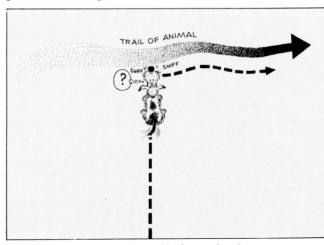

TRAIL OF ANIMAL

How the dog detects the trail of an animal.

The Talking Animal dogs.

for up to five days after the meal. This mechanism is invaluable to wild dogs. They eat, taking in the butyric acid, and then in a few days when they are hungry once more the chemical has prepared their noses for the next hunt. At such times the sense of smell may be three times more acute than normal.

Hearing is something else at which dogs are excellent. They are equipped with large external ears that are served by 17 muscles. They can prick and swivel the sound receivers to focus on the source of any noise. They can hear sounds which human beings cannot, and are sensitive enough to be able to tell the difference between two metronomes, one ticking at 100 beats a minute and the other at 96 beats a minute. Dogs can shut off their inner ear in order to filter from the general din those sounds on which they want to concentrate.

Like cats, dogs are very sensitive to vibrations and will give warning of earth tremors some time before humans are aware of any movement in the ground.

People often say that one year of a dog's life is equivalent to seven of a man's. This is not true. A more realistic approach has been worked out by a French veterinary surgeon who suggests that the first year of a dog's life equals 15 human years, the second equals a further nine human years and thereafter each dog year counts for four human years. This provides us with a table:

Age of dog	Equivalent age of man
1 year	15 years
2 years	24 years
3 years	28 years
4 years	32 years
8 years	48 years
12 years	64 years
15 years	76 years
20 years	96 years

Like the owners, dogs are living longer nowadays, but few pass 17 years (84 human years). The record for canine age is claimed for a 27¼-year-old black Labrador that died in 1963 in Boston, Lincolnshire, though there are less reliable reports of another dog tottering up to an incredible 34!

5. The Hedgehog

The hedgehog is an insectivore, a fairly primitive mammal, 17 species of which are distributed across Europe, Africa and Asia as far north as deciduous woods grow.

The back and sides of its body skin are covered in spines – about 6,000 in total on an average hedgehog. The spines are a highly specialised defence mechanism. The spines can also act as shock absorbers when a climbing hedgehog (they climb quite efficiently) descends by just dropping off the top of the wall, log or whatever it is standing upon.

Hedgehogs are highly resistant to many poisons and can survive the bites of snakes that would kill ten men. They can gobble bees and wasps without being troubled by the insects' stings.

Hedgehogs are the noisiest British mammals with a range of sounds that includes snorting, hissing, coughing, cackling, puffing, grunting and screaming.

Baby hedgehogs are born between May and July with a second litter sometimes produced during August and September, after a pregnancy period of 30 to 40 days. There are normally three to seven blind and deaf babies in a litter, each weighing about nine grams. The spines of a newborn baby hedgehog are pale, soft and rubbery and are flattened into the skin which is specially soggy with a high water content. One and a half to three days after birth, a second layer of spines begins to grow through the first spines that are now standing upright. A baby cannot roll into a ball until it is about ten days old.

Hedgehogs have moderately good eyesight and some scientists think they see the world in shades of one colour only – yellow! They have excellent hearing and a marvellous sense of smell. When a hedgehog starts sniffing about for food his nose will start to run – the moisture increases the nostril's ability to pick up smell molecules in the air.

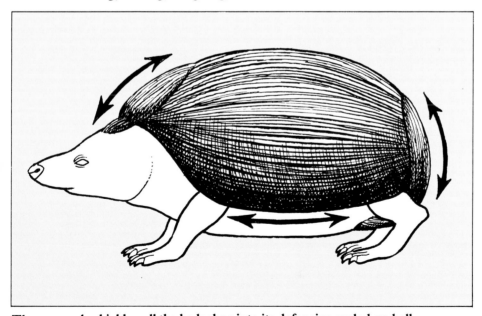

These muscle shields pull the hedgehog into its defensive curled-up ball.

Jennifer, the Talking Animal hedgehog, disproves the legend that these animals gather fruit on their spines.

Hedgehogs tend to be solitary and will fight with other hedgehogs when they meet them. They will even 'fight' brushes and brooms that *look* like other hedgehogs! The main enemy of the hedgehog is man and his motorcar. Many hedgehogs 'freeze' when crossing the road at the approach of a vehicle and get run over. Few animals can tackle the hedgehog with the hope of winning. Large birds of prey, badgers, polecats and martens are sometimes successful. Foxes and dogs do not stand much of a chance.

The hedgehog, sometimes called 'hotchi-witchi' or 'urchin' by countryfolk, is the subject of many quaint beliefs and legends. It is said that the animal will steal milk from cattle, thieve hens' eggs and collect fallen fruit in orchards and gardens by rolling onto them and impaling them on its spines. The hedgehog does occasionally lap milk drops from the teats of a cow that is lying down, and it may enter hen-houses in search of shelter and food and be tempted to sample the contents oozing from an already cracked egg. I doubt if it ever takes and cracks hens' eggs itself. The story of the fruit-gathering is certainly nothing more than a charming myth.

Although classified as an insectivore, the hedgehog has a broad (omnivorous) diet. It particularly likes earthworms, but also preys on insects, woodlice, spiders, snails, frogs, toads, lizards and snakes. They sometimes eat young birds and mammals, and when food is scarcer will take mushrooms, nuts, fallen fruit and berries. If you are lucky enough to have a hedgehog visit your home foraging for food, put out things like tinned cat or dog food, dog biscuits, hard-boiled egg and cheese. You can also offer 'desserts' like rice pudding, milk pudding and blancmange! It is sensible to add a few drops of multi-vitamin syrup to each meal.

Hedgehogs are fine little swimmers but do not generally take to the water unless obliged to.

During winter (October to April), hedgehogs hibernate in a snug place – a heap of dried leaves or a hole in a bank that they line with moss and leaves. While hibernating, the animal lets its temperature drop and its heart and breathing rates slow significantly. It builds up energy while it sleeps from a special brown fat stored beneath the skin of its back. The brown fat, which can release heat 20 times faster than ordinary fat, acts as a sort of thermostatically-controlled electric blanket. The colder the surrounding temperature, the more heat the fat releases to the hedgehog's body.

Sometimes, when it comes across a particularly pungent smelling substance – for example boot polish, a cigarette end or an old sock – a hedgehog will become very excited, start to slaver at the mouth, lick the saliva into a froth and then spread the froth over its spines. This so-called 'self-anointing' is a strange piece of behaviour whose significance is not yet understood.

A hedgehog feast.

Hedgehogs are fine little swimmers – but do not generally take to the water unless obliged to.

6. The Giraffe

The giraffe was called 'camelopard' in mediaeval times – folk thought it was a cross between a camel and leopard. The giraffe lives in the savannahs and open woodlands of Africa to the south of the Sahara desert. There are nine sub-species who have variations in basic coat patterns. Every individual giraffe has its own unique coat pattern, as personal as fingerprints in human beings. It can grow up to almost six metres high, lives for 25 years in the wild and slightly longer (28 years or so) in wildlife parks and zoos.

Giraffes have a peculiar way of walking. They pace – moving both legs on the same side forward at the same time. When galloping, the hind legs are brought together more or less at the same time and are set down outside the forelegs. They can travel at up to 65 kilometres an hour. The long legs and heavy hoofs of a giraffe also serve as fearsome weapons. They kick powerfully in all directions and can easily kill a lion with one blow.

The eyes of a giraffe can detect small movements at a distance of three kilometres and the field of vision is better than a cinemascope screen at almost 270 degrees. Their horizon-scanning ability is largely due to the possession of a horizontal letterbox-shaped pupil (not round as in the human eye).

Giraffes are ruminants (cud-chewing animals like cows), and unlike most such creatures are born with horns. The horns, still rather rubbery, lie flat on the head of a newborn baby giraffe but stand up by the end of the first week of life. Giraffes are browsers – they love the leaves and shoots of bushes and trees, particularly those of acacias. They can easily cope with thorny plants. They do this by means of the heavily grooved and thickened roofs of their

Talking Animal giraffes at Longleat.

Male giraffes stretch their necks up into the trees to eat, *female* giraffes bend their necks to eat.

Giraffes always leave a couple of friends on guard when drinking.

mouths and by production of large quantities of tacky saliva that protects the softer parts like the tongue. The giraffe's front teeth are broad and splayed, with lobed 'fang' or canine teeth specially for stripping leaves off branches. There are no teeth at the front of the upper jaw – just a hard pad that the lower front teeth bite against when neatly breaking off non-spikey shoots and leaves. The black tongue is very long, mobile and sensitive. Giraffes dine for over twelve hours every day, mainly in the twilight, at dawn and dusk. If there is plenty of moonlight they adore midnight snacks too!

Apart from the lion and occasionally the leopard, the giraffe's principal enemy is man. The latter has hunted it for meat, and even just to take the tail hairs to make bracelets for the tourist trade. The tuft of tail hair serves the giraffe as an efficient fly whisk.

But they are at their most vulnerable to attack – with their heads between their legs.

The giraffe has excellent senses of smell, hearing and sight. Its vision is particularly acute. The animal's height gives it an ability to act as a living watchtower – that is why other creates such as cattle and wildebeest like to graze near giraffes. Their long-necked friends act as early warning devices, detecting the approach of predators. When some giraffes are at their most vulnerable with legs splayed drinking, they always make sure to have one or two of their friends acting as sentries with heads held high!

Giraffe mothers give birth after a pregnancy of about 15 months. The females give birth in giraffe 'maternity hospitals' and the same calving grounds are used year after year. One female may give birth to a dozen babies during her lifetime.

Bull giraffes occasionally use their horns for fighting but generally the combat is not serious, more a sort of ritual sparring involving the intertwining of their long necks – a behaviour appropriately called 'necking'.

The long neck of the giraffe does *not* as you might imagine contain far more neck bones than is found in a short-necked mammal like man. In fact the number of neck bones is the same in all mammals – seven – except for the sloth and the manatee or sea cow, which have six. Giraffe neck bones are simply remarkably long and dwarf those of a human or even a horse.

How does blood get up to the brain of a giraffe? The blood pressure of a giraffe is higher than that of, say, a cow. Also the blood vessels in the neck have very elastic walls and there are special valves in the neck veins to assist the blood flow.

The giraffe's only living relative is the rare and mysterious okapi, an animal first discovered by western scientists in 1901 in Zaire. A deep forest animal, the okapi resembles the giraffe in having a long neck, a similar shaped head, peglike horns, a long black tongue and some light patterning on the legs. It sports a gorgeous velvety deep-chestnut coat.

Babies are born in giraffe hospitals. When it is born the baby giraffe can drop 2 metres from its mother to the ground without being hurt.

7. The Camel

There are two types of camel. The one-humped camel or Dromedary is a native of the hot desert of south-west Asia and north Africa. The two-humped camel or Bactrian hails from the steppe-lands of Mongolia. The camel, it has been said, is 'a horse designed by a committee'. I do not agree. It is one of the cleverest, most ingeniously designed animals I know.

Here we have an animal that can survive and work in the harshest and most arid conditions. It can go as long as 21 days without water and cover 900 kilometres during that time. How does it do it? Not by storing water in its hump, its three stomachs or anywhere else in its body. The camel's hump is actually composed of concentrated fat, a useful source of energy when food is in short supply. The camel carries it neatly in a 'knapsack' on its back rather than as a layer evenly distributed over its body beneath the skin (as in man and other animals) where it would interfere with the easy dispersal of excess heat, like wearing a tight sweater!

Nomads gather at Tan Tan in the Sahara desert.

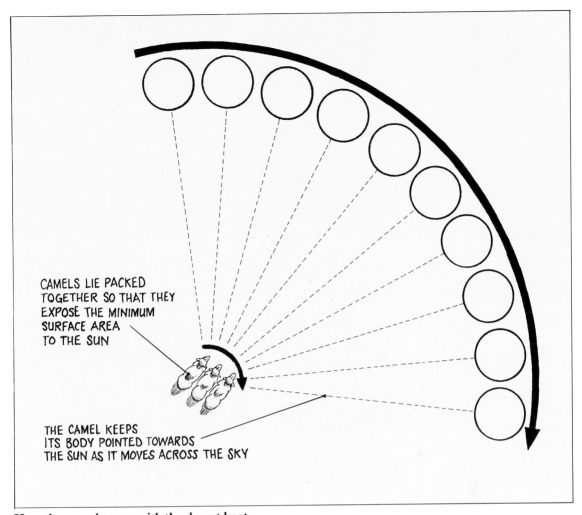

How the camel copes with the desert heat.

The camel copes with heat by various mechanisms:

1 By letting its body temperature rise several degrees above normal during the hottest part of the day without getting a fever or damaging its brain as would happen in humans, and then letting the body temperature fall below normal at night. It behaves as a sort of storage heater, taking in heat during the day and radiating during the cool desert nights.

2 It lies down early in the morning and keeps the same patch of cool sand beneath it throughout the day if at all possible.

3 It keeps its body pointed towards the sun as it moves across the sky with its legs tucked neatly underneath it. This exposes the minimum of body surface to the fierce source of heat.

4 Camels lie packed together on the ground so that they expose, as a group, the minimum surface area to the sun. They take in less heat from their pals than they would do from the sun.

The camel's survival kit.

The camel's coat of hair acts as an efficient heat barrier or insulating layer. It not only keeps the animal warm during the nights but also keeps outside heat from getting in during the day. Human beings in hot climates (Arab bedouins, for example) tend to do the same thing. They cover themselves in flowing garments to protect themselves from the heat. Camels do sweat, but only when the temperature gets very high and then only enough to produce a cooling evaporation through the coat and not enough to wet the hair which would reduce heat loss from the body.

Camels are very careful not to lose water unnecessarily. Indeed they are miserly about it. They hardly ever pant (which loses water) unlike cattle, dogs, man and many other animals. Their kidneys produce only small quantities of concentrated urine in order to conserve water and they can take back into the body waste products such as urea for conversion into food, which apart from being economic also cuts down the need to lose water. They are also one of the few mammals that can drink very salty water such as seawater without any trouble.

Camels can lose a third of their body weight when water is scarce and their tissues begin to dry up. Human beings in a similar situation would lose water to the point where their blood begins to thicken, circulation problems develop and death from heat-shock quickly follows. Even when the camel's body tissues are being squeezed dry of liquid, the blood stays thin. When a camel does eventually reach water, it can regain weight amazingly quickly by drinking non-stop for ten minutes and taking on board 150 litres of water. I often think that the camel in the desert can be regarded as a sort of living sponge!

Camels possess glamorous long eyelashes and slit nostrils that can close tightly to keep out windblown sand. Their broad two-toed feet distribute their weight over a large area and enable them to walk with ease over soft sand.

The broad two-toed foot of the camel distributes its weight so that walking on soft sand is easy.

They are unfortunately rather grumpy creatures and often 'spit'. This is a mild defence mechanism and is actually quite unlike human spitting. Rather it is the bringing back of large quantities of digesting stomach contents which are then sprayed accurately in any direction, generally towards the person who has upset them. Other members of the camel family such as llamas have the same rude habit.

Arab camel handlers sense when a particular animal is getting uptight. To avoid being injured (camels can inflict terrible bites), they take off a garment and throw it down in front of the animal which then proceeds to work out its frustrations by jumping up and down on, and biting the material which acts as a 'substitute person'. After a few minutes of this the camel feels much relieved and calmer and once again co-operates uncomplainingly with its handler.

The male camel often 'blows bubbles' by puffing out a bag of membrane that lines its mouth – the dulaa. This strange habit is actually part of the normal mating behaviour, a sort of showing off in front of other males.

Camels are very useful to man in a variety of ways. They have been used for meat, milk, leather and wool, as transport, beasts of burden and a form of money, and for racing, fighting displays and religious ceremonies. As a baggage animal a camel can carry up to 200 kilograms for 33 kilometres per day. It is a good strong swimmer and has been seen to cross the Nile at points where the stream is very broad and the current powerful. It is however, unable to jump, and ditches, even narrow ones, can present serious obstacles to the camel unless it is able to stride across them. The body of the dromedary camel is conveniently protected by patches of thickened rough skin on the elbows, knees and breastbone when lying down.

Long eyelashes and nostrils that can be tightly shut to protect the camel against the sun.

8. The Cow

Cattle belong to the bovine tribe of uminating (cud-chewing) hoofed animals. This tribe includes the domestic cow, the yak of Tibet, the American bison and the very rare Cambodian forest ox or Kouprey, only a handful of which may still survive. The modern cow evolved from a group of antelope-like animals, some of which survive today in the form of the Nilgai and the four-horned antelope of India.

land. These animals have reddish ears, black muzzles and black-tipped horns, and sometimes give birth to a black calf. Traditionally such births were thought by countryfolk to be predictions of a death in the family of the owners. It certainly seemed to be true until the mid-19th century, as whenever a black calf was born a family death followed. Since about 1842 the grim omen seems to have lost its power as black calves have become more common.

Cattle have good hearing and sight and are sensitive in taste and touch,

Aurochs, ancestor of the domestic cow.

During the process of evolution a most successful ancestor of the cow arose in Asia and spread out to populate much of the world. It was the Aurochs, the last one of which died in the seventeenth century. From the Aurochs various kinds of wild cattle developed and in England some still survive to this day in the form of the White cattle which can be seen at Chillingham in Northumber-

particularly around the lips, but they rely mainly on their excellent sense of smell to detect the presence of enemies. They can see some colours, but not red. The red cape of the Spanish matador is charged by the bull not because it is red but because it looks bigger than the slim outline of the man's body. The bull, poor thing, thinks the cape is where most of the man is!

White cattle at Chillingham, Northumberland.

Cattle are herd animals and it was their social organization which made it easy for early man to domesticate them. Cattle may seem to be rather colourless and placid creatures, but some of them are among the most dangerous animals in the world. African buffaloes are often said by experts to be the most dangerous animals of that continent – more

deadly than lions, leopards, crocodiles or cobras. Great numbers of people have been killed by the African buffalo, which is a tough and fearless creature.

The interesting thing about the cow is the fact that it is really a living chemical factory or, looked at another way, a living zoo full of hard-working cheerful germs! Unlike humans, dogs and cats, who all break down the food they eat by means of stomach acid and other chemicals, cattle use millions of friendly little germs in their innards to digest the food for them. The germs have the knack of breaking down the tough cellulose which surrounds the more nutritious parts of plants. This means that cows can eat grass, hay and other vegetation which man's digestive system simply cannot handle.

The cow has four stomachs and keeps its teeming millions of digesting bugs in a sort of soup within the first stomach, or rumen. The germs wait for the grass to arrive down the throat and then set to work. The digesting food eventually passes

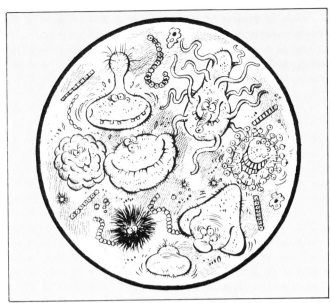

Hard-working, friendly germs live in the cow's stomach.

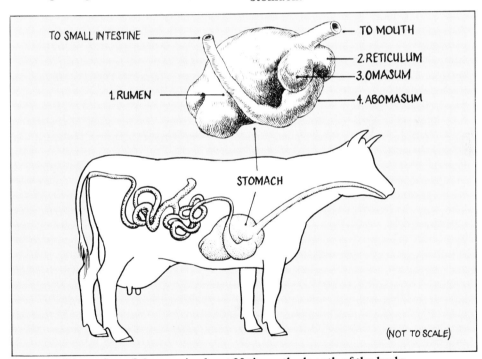

The small intestine of the cow is about 20 times the length of the body.

on to the second stomach, or reticulum, which is lined rather like a honeycomb. Bowls of food (cud) are prepared there and then pass back up to the mouth by way of a special by-pass tube for further chewing. The cud now goes back down to the second stomach and passes to the third stomach, or omasum, which is packed full of layers of tissue like the leaves of a book. Here water is pressed out of the digesting cudbowls. At last, the food goes on again to the fourth and final stomach, or abomasum, which more closely resembles a human stomach.

The chewing of cud in this way probably developed because it allows a cow to grab as much food as it can when it is available. It can then hide out of the way of predators while digesting it at leisure.

The big mass of digesting food in the stomach of the cow contrasts with the situation in the horse where digestion by friendly germs also takes place, this time not in the stomach but in the large intestine. The main weight of food is thus farther back in the horse than in the cow. That may explain why horses on rising get up on to their front legs first and then heave up their back end while cows do it the opposite way.

With its soup of digesting food and busy germs, the cow's stomach is just like a fermentation vat or tank in a brewery. As in all such chemical plants, heat is produced. The cow thus has a useful internal central heating system which allows it to cope well with cold weather. Like the process of fermentation, digestion in the cow produces gas. So cows burp quite naturally from time to time. If they cannot burp for any reason, they blow up with gas – a condition called bloat – and the vet has to be called to release the pressure.

Milk production is a key feature of all mammals. It is a perfect food for baby animals. The cow's natural production of milk has been increased enormously by means of

The horse, with its main weight to the back, gets up front legs first.

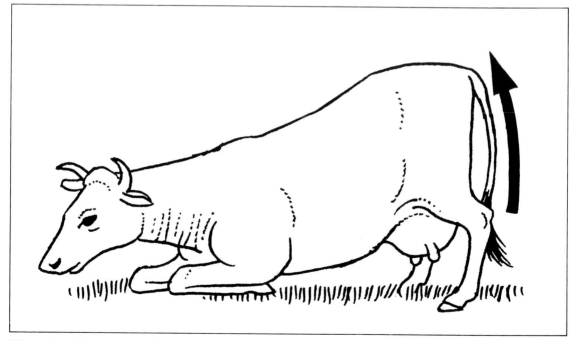

The cow, with its main weight further forward, gets up hind legs first.

Andrea milks Betsy, the day after she gave birth to a delightful bull calf.

selective breeding and the techniques of modern science. Cow's milk is not the richest mammalian milk by any means. It contains only about 4 per cent fat and 3.5 per cent protein. By comparison, blue whale milk contains 42 per cent fat and 11.7 per cent protein, but you cannot milk blue whales!

Milk is produced by the cow when a calf is born at the end of a pregnancy period that lasts around nine months. The first milk (colostrum) as in all mammals is unusually rich and contains lots of chemicals called antibodies that protect the calf against a range of diseases. A prize dairy cow can give as much as 85 litres of milk in a day and 20,000 litres in a year.

9. The Dolphin

Everyone loves the dolphin. Probably the second most intelligent animal in the world, it is a delightful cheerful whiskers. They fall out during the first few weeks of life, but the follicles where the whiskers sprouted can be seen even in old dolphins on each side of the upper lip.

Dolphins have big eyes which, because of a spherical lens instead of an ovoid one as in humans, can see as well out of as in water. They produce

The Common dolphin.

character which has superbly mastered life in the oceans.

The dolphin is not a fish but a mammal. Its ancestors once walked on earth. Unlike a fish, it has no gills but breathes air through lungs. Its flippers are built completely differently from fishy fins (there are no bones in a fin, but in a dolphin's front flippers you can find all the bones of the human hand), it is warm-blooded, not cold like a fish, and it has hair – a characteristic of mammals. Again like all mammals, the mother feeds her young on milk. The hair on a dolphin is seen now only in the newborn dolphin as

large syrupy tears, not because they are sad, but to protect their eyes against the friction of salt water as they swim.

They can dive deep and stay under water for long periods. Some of their close relatives, the whales, can stay below for two hours. Dolphins and whales do this by taking down with them reserves of oxygen and, more importantly, switching off those body processes that need oxygen while under water. When they return to the surface they don't suffer from a disease called 'the bends' like human divers. The blood system is cleverly designed to avoid this.

Scientists think that the dolphin evolved from a sheeplike land animal that lived many millions of years ago. It began to mess about in the shallow inshore waters of the sea – it felt safe from predators when paddling! It enjoyed swimming more and more, and found over thousands of years that fish tasted even better than grass.

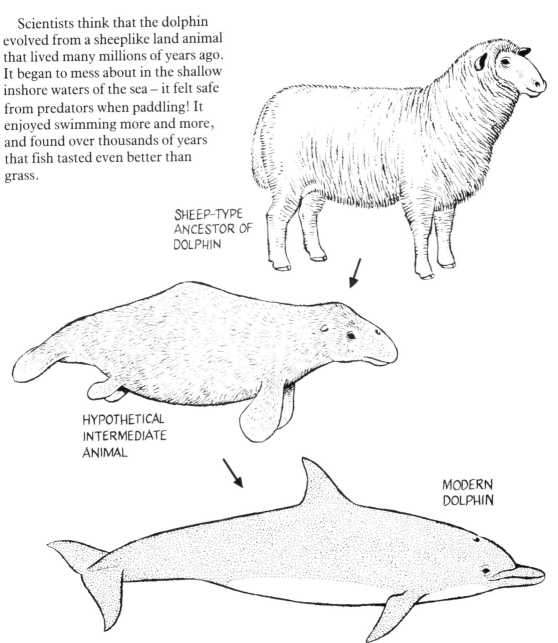

SHEEP-TYPE ANCESTOR OF DOLPHIN

HYPOTHETICAL INTERMEDIATE ANIMAL

MODERN DOLPHIN

Slowly, the dolphin became increasingly adapted to a totally aquatic life. It lost its hind legs (although you can still find remains of hindleg bones deep in a dolphin's body) and it enlarged its tail into a paddle. Its front legs became flipperlike but retained the front leg bones within them. Its hair or wool disappeared (except for the whiskers) and its nostril moved steadily up its face from the tip where you find it in

Dolphins may be descended from sheep.

land mammals to the top of its head where it became the dolphin's blowhole. The dolphin did however keep some bits of its sheepish ancestor such as the multi-chambered stomach, the kidney, certain typical blood proteins and, unfortunately, a tendency to some of the diseases of sheep.

45

Dolphins have a marvellous sense of hearing and use sound beams for communication and also echo location (sonar) in dark, deep water. They send out streams of clicking noises from their voice-boxes (larynxes) which are focused by a 'sound lens' made of special fat (the round part of a dolphin's forehead that you can see). The sound beams bounce off objects at great distances and are received by the dolphin on the tip of its chin and at its armpits! The sounds then travel along special pathways to the very sensitive inner ear.

There is also a tiny outer ear which you can see if you look closely at a dolphin – it is a dark spot as big as a match-head behind the eye. Using their sonar, dolphins can tell the range and speed of an object and also its nature. They can distinguish between for example a herring and a mackerel and between a shark and another dolphin. It seems that they can tell one fishing boat from another. Human scientists have yet to build sonar machines for submarines that are half as good as the dolphin's system. Study of the clever dolphin has helped mankind to design mechanical aids for handicapped blind and deaf people.

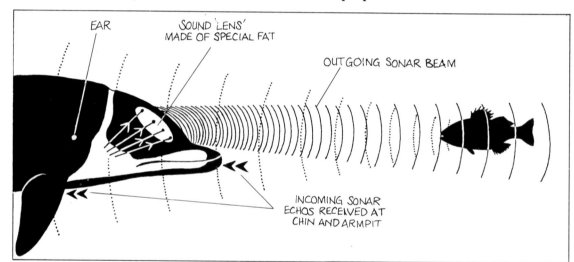

EAR

SOUND 'LENS' MADE OF SPECIAL FAT

OUTGOING SONAR BEAM

INCOMING SONAR ECHOS RECEIVED AT CHIN AND ARMPIT

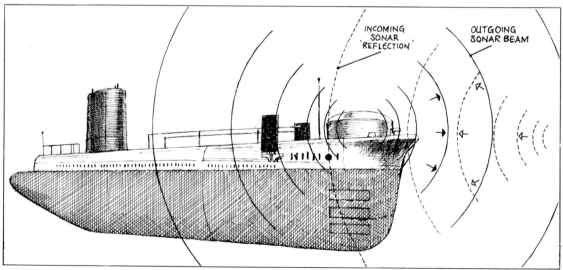

INCOMING SONAR REFLECTION

OUTGOING SONAR BEAM

Dolphins have better sonar systems than submarines.

David's favourite dolphin, Commerson's dolphin, in waters round South America.

Dolphins talk to one another in a language which has been extensively recorded and analyzed by computers but which no one has yet been able to translate. There is evidence to suggest that they can tell the emotions and feelings of another dolphin by looking into its head with a sonar beam and detecting changes such as blushing in the lining of spaces (sinuses) within the skull.

Dolphins swim very efficiently and go faster through the water than scientists could explain for many years. It is not just that their bodies are streamlined and smooth. Now we know that as they accelerate they literally jump out of their skins. They leave behind a 'ghost' image of themselves, a layer of skin one cell thick, on which water drag and turbulence works. This allows them to shoot away unhindered rather like a bag of soap popping out of the grip of a slippery hand.

A dolphin baby is born after a pregnancy of one year. It suckles from its mother under water for a

David holds the 'beak' of Lulu, the dolphin, while applying covers to her eyes before showing her marvellous ability to see 'blind'.

David taking a blood sample from a dolphin at Windsor as part of its regular check-up.

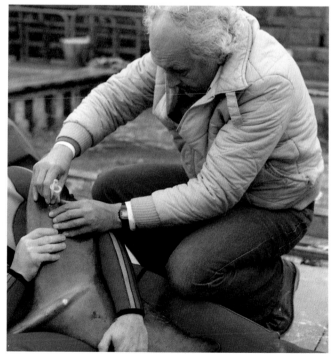

further one to two years. The dolphin mother produces very rich milk which she pumps quickly into the baby's mouth to avoid it having to suckle too long submerged. When a dolphin gives birth and while the baby is young, another female dolphin acts as a 'nurse' or 'auntie' and shares the work of looking after and teaching the infant. You can see baby dolphins that have been born in a marineland in many parts of Europe including Windsor in England. Dolphins can probably live for twenty-five to thirty years.

10. The Snake

Most people think that snakes are unpleasant, slimy to the touch, aggressive, rapidly slithering creatures that sting with their fork tongues. This is not true.

Snakes are reptiles, members of a family containing over two thousand living species. They are air-breathing, cold-blooded animals with bony backbones (vertebrae), whose skin is covered with protective scales and who produce young either alive or in eggs.

Snakes do not have a larynx (voice-box) or vocal cords, so they cannot produce a true voice. Their hiss is made by the expulsion of air from the lungs. In some snakes the air passes over a gristly 'tuning fork' at the entrance to the windpipe which produces a slightly musical note.

They have well-developed vision, but are mainly sensitive to moving objects. Stand absolutely still and a snake may well not notice you. The eye of the snake is completely covered by a transparent layer of skin. This is normally shed along with the rest of the outer layers of skin when the snake 'sloughs' periodically as it grows. Rattlesnakes do not slough all of their skin. They keep a bit from each at the tip of the tail with which they make their rattles to serve as a warning device.

Snakes move not in caterpillar fashion up and down, but in side to side waves of muscular contraction that run down the body from head to tail. The contractions act upon the skin scales beneath the body which are attached to the ends of the ribs and point backwards. The scales get a grip on the ground and propel the animal along. You could say that a snake walks on its ribs. A snake on a perfectly smooth surface such as a glass plate will get nowhere at all. Snakes look as if they move fast, but this is mainly an illusion produced by the waves of contraction running down their body. In actual fact 11 kph seems to be about the fastest they ever go although 24 kph may be possible downhill when they are fleeing from something like a forest fire.

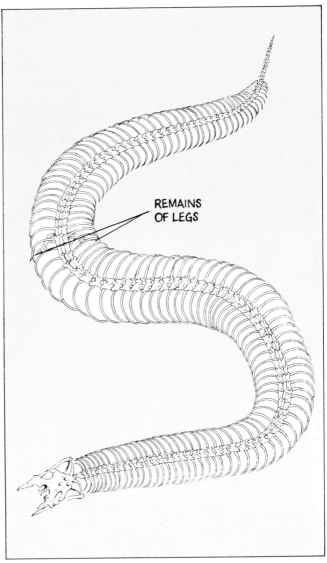

REMAINS OF LEGS

The snake is descended from reptiles who possessed legs – the remains of hind legs can still be found in the body tissue of some snakes and certain species show two small claws, all that remain of the hind feet, on the outside of their belly skin. They are really limbless lizards.

49

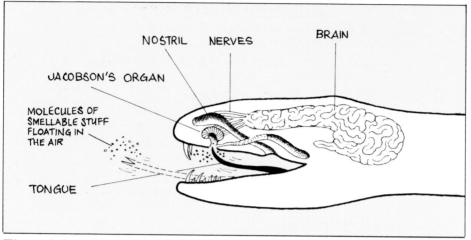

The snake's moist tongue picks up molecules of smellable substances hanging in the air, brings them back into the mouth and touches them to a sensitive spot (Jacobson's organ) in the roof of the mouth. This sends nerve messages to the brain.

Snakes can be found in all parts of the world except the polar regions, New Zealand, Ireland (legend is that St Patrick drove them away), the Azores and most of Polynesia.

The hearing of snakes is different to that of mammals. It appears they 'hear' not through an external earhole but by picking up vibrations from the ground which are magnified by their long single lung and are transmitted to their inner ear. A snake listens with its chest.

Snakes can smell quite well through their nostrils, but they also have a marvellous way of smelling or tasting with their forked tongue.

When they flick their tongue in and out they are actually tasting the air ahead of them.

Snakes such as rattlesnakes are able to see in the dark by detecting infra red heat radiation. Between the nostril and the eye there is a heat sensitive 'eye' lying in a tiny pit. The snake can use this device not only to locate prey in total darkness but also to hunt them by day even if they camouflage themselves. The prey's body heat gives them away.

Snakes are not strictly *cold*-blooded. Their body temperature is that of their environment. In other words when their surroundings are

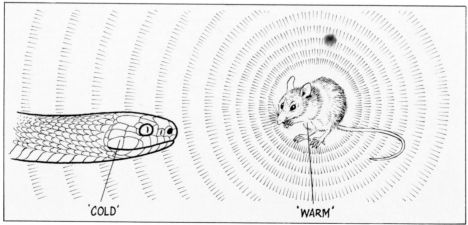

The body heat of the snake's prey gives it away.

cold, they get cold and less active. When it warms up, they warm up and become more active.

Snakes do not go looking for trouble but use their weapons only when on the defensive or in search of a meal. Some snakes are poisonous. The poison or venom is made in glands (modified saliva glands) lying in their heads, and it is injected into their victim through hollow-fanged teeth. There are two main kinds of snake poison – one affects the nervous system and produces death by depressing the brain, the other affects the blood system and damages the blood cells and the blood vessels. Antidotes to snake poisons are available to doctors and vets and are called antivenins.

It is possible to 'milk' snakes without hurting them in order to obtain some of their poison, which can then be studied by scientists who have found ways of making several useful human and animal medicines from them.

The most poisonous snakes in the world come from tropical countries. Experts agree that the most dangerous species is the King Cobra of India, Malaysia, Burma and Indo-china. It is a big, bad-tempered and vigorous snake that injects a large quantity of highly poisonous nerve-damaging poison into its victim. Luckily the King Cobra is normally found only in deep jungle. My list of the ten most dangerous snakes in the world is as follows:

King Cobra

Mamba

Tiger snake

Puff Adder

Death Adder

Diamondback Rattlesnake

Russell's Viper

Bushmaster

Ringhals

Indian Krait

David prepares to grab a cobra prior to 'milking'. To 'milk' a snake it must be held firmly behind the head with its mouth open. Its fangs are then brought close to a rubber membrane stretched across the mouth of a glass beaker. The fangs are quickly pricked into the membrane. To the snake it feels as if it has just pierced an animal's skin and it automatically squirts out its poison into the beaker.

But there are many more runners-up. Venomous snakes kill more people in a *year* than sharks do in a whole century, more people in a *month* than big cats such as the tiger and lion do in a century and more people in a *day* than rogue elephants have done since the beginning of time! In India alone, 20,000 people die every year from snake bites.

The only poisonous snake in Great Britain is the adder.

Although snakes are said to strike and bite faster than the eye can see, in fact the human hand and the mongoose (which relies on its speed and nimbleness alone to attack and eat venomous snakes without getting injured) are faster.

Some snakes are not poisonous but kill their prey by constriction. This does not mean that they strangle their victims. What happens is that they usually first bite their victim and then throw a coil or two round its chest. When the victim breathes in, the snake 'takes up the slack'. The victim has less room still in which to breathe, so it pulls its chest in still further at the next breath. Again, the snake takes up the slack. Eventually the victim has no more room left in which to breathe. In a sense, it has suffocated itself.

There are only three or four authenticated cases on record of humans who have been killed and eaten by constrictor snakes. Most constrictors take nothing bigger than rabbits or small deer. The biggest animal ever eaten by such a snake was a 60-kilogram impala antelope.

A rock python devouring a gazelle.

11. The Horse

Before I begin, here is a quiz. Try it yourself and on your friends. I bet you cannot say who owned each of these famous nags;

Dapple
Incitatus
Bucephalus
Copenhagen
Agnes
White Surrey
Black Bess

And who did the winged horse, Borak, carry?

The answers are on page 63.

The horse family, which includes horses, asses and zebras, has a long history of evolution. About 55 million years ago in what are now called Europe and North America but were then joined together in one land mass, a little doglike animal moved through the forests browsing on low shrubs. This was *Hyracotherium*, ancestor of the horse and also of the rhinoceros and tapir. Descended itself from forebears possessing the basic five toes on each foot, it had already lost two outer toes on its hind feet and one inner toe on its forefeet, the remaining toes looking doglike with pads. There was no sign of hooves as yet. It had a short muzzle and a long tail held curved like a cat's.

As millions of years passed, the descendants of *Hyracotherium* split off into numerous branches of the family tree. The rhinoceros and the tapir went their way; the horse branch went theirs, discarding more toes so that they might concentrate on the perfection of a single highly efficient toe on each foot, growing a larger body, developing teeth that were ideal for cropping and grinding grass, and acquiring large efficient eyes.

Domestication began around 3000 BC almost certainly in Eastern Europe and on the Steppes of western Asia, probably by reindeer herdsmen. All 'wild' ponies (excluding the Przewalski Horse) which exist today and including the nine British breeds, should really be called 'feral'

Horse and man (in full fifteenth-century battle armour).

– descendants of domesticated stock which went back to the wild. Some experts consider that certain breeds of British pony *might* have some truly wild blood in them. The Exmoor just might be a true pure-blooded wild pony inhabiting a part of Britain that escaped the glaciers of the Ice Age, and there is speculation too about the Connemara.

The horse is built as we find it today essentially because of a fundamental change in the earth's climate that began about 24 million years ago. The weather became more distinctly and regularly seasonal, it was drier and the tropical zone contracted towards the Equator. Forests began to retreat and everywhere the plants we call grasses became abundant. Open savannahs of grassland were established and, to take advantage of them, the horse descendants of *Hyracotherium* moved out of the bush. So that these creatures could cope with the benefit and dangers of the new wide open spaces, evolution changed the structure of the equine body in various very sensible ways.

Przewalski's wild horse from the Altai Mountains of Mongolia.

Long nose, good eyes, sharp ears and legs designed for speed provided the main defences for the grass eating machine when it moved out of the forest.

The three key requirements for the plains were: 1. Efficient tools for cropping and grinding grass which, unlike the leaves on which its forest-living ancestors had lived, contained much harsh silica as well as more tough fibre. 2. Mechanisms for detecting enemies at a safe distance, and 3. Ways of avoiding those enemies now that the protective cover of the trees had been left behind.

The horse was well equipped for doing these things. It has two excellent types of teeth for dealing with grass – sharp incisor teeth for cropping it and special high crowned cheek teeth for grinding it. The horse has a relatively simple stomach more like man's than a cow's, and it has very big large intestines in which millions of germs aid the digestion of grass fibre.

Open spaces led to the development of large eyes set on the sides of the head, perfectly positioned to watch the horizon for any hint of danger whilst grazing. Excellent hearing with mobile, 'direction-finding' ears helps in the same way. The horse has very good sight and can spot a man at a distance of almost two kilometres. As the eyes are set at the sides, the fields of vision do not overlap and so they are unable to see in 3D as we do. On the other hand, the horizontal letterbox-shaped pupil in the eye gives a far wider visual sweep of the distance than a round pupil would.

Although it cannot see as much of the colour spectrum as man and other primates, the horse does have a wide range of colour vision and in this respect are on a par with the giraffe, nsheep and squirrel. It can see much fine detail; its cousin the zebra can recognize one another by the particular pattern of their stripes. No two zebras have identical striping.

The mobile ears are used not just to catch sounds but also, particularly in wild horses, as a means of communication. They wag them to one another in a kind of simple semaphore.

The sense of smell is first class. The nostrils are large and supple and can be closed tight against dust or sandstorms. In a wind a wild ass can scent a man at 500 metres.

The horse has very good sight and can spot a person almost two kilometres away.

A power and speed machine.

Horses have a larger body size than their ancient forest dwelling ancestors; bulkier bodies both use food and conserve heat more efficiently. To out-distance any attackers, the horse family concentrated the immense power of their body muscles into just four toes – one on each leg. The other toes were largely discarded so that the one which was left could be specially modified for superb running by turning the primitive nail into a hoof and devising a complex system of ligaments in the fetlock. Side to side movements decreased and so gave the leg much extra forward spring.

Wild horses and their relatives can be fairly nippy movers. The wild ass is particularly swift and untiring, and some experts consider it superior in many respects to riding horses. Onagers have been timed running at an average of 45 kph for nearly half an hour with burst of over 50 kph. Compare this with a record for a thoroughbred racehorse under ideal conditions over five kilometres of 55.1 kph. Even over a short distance the racehorse is unlikely to surpass 65 kph. Wild asses are no mean jumpers either: onagers have been seen to clear walls 2.3 metres high.

Although wild horses do not live much beyond 20 or 25 years, with the occasional individual reaching the low 30s, the age record for a domestic horse is no less than 61 years!

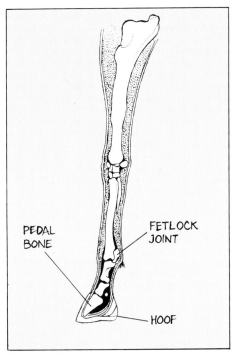

PEDAL BONE

FETLOCK JOINT

HOOF

The foreleg and hoof of a horse: a perfect single highly efficient toe on each foot.

12. The Bird

If you go to the back door right now and take a look out, I can almost guarantee that you will see, going along the street or perhaps sitting on the garden wall – a *dinosaur!* For the probability is that all of the nearly 9,000 species of living birds known to science are in fact dinosaurs that took to the air. Look at that humble little sparrow and remember *Tyrannosaurus rex!*

One hundred and forty million years ago when dinosaurs ruled the earth, some of them were quite small but were still proper dinosaurs. They ran over the ground on their hind legs and may have done a bit of shinning up trees. Ah, but you say, dinosaurs were reptiles – cold-blooded animals – and birds are hot-blooded with body temperatures of 40°C. Could they in any way be related? Yes, there is plenty of modern scientific evidence, including the belief that dinosaurs were warm-blooded.

A tiny dinosaur, 60 centimetres high, ran over the ground on its hind legs.

The first dinosaur/bird was called *Archaeopteryx* – his fossil has been found in Germany. In many ways he was still a reptile, for he had teeth instead of a beak, scales over part of his body and bones in his long tail (modern birds do not have any bones in their tail feathers). But the fascinating thing is that he had *feathers* and it is feathers that make a bird.

He also had hollow bones to reduce weight and one of his toes pointed backwards – a useful device for perching on branches. Probably these first birds could not take off from the ground, but had to climb into the trees and launch themselves into the air to get going. They would clamber about the branches still using the claws present on their forelegs (which were now wings). There is a rare living bird, the hoatzin in South America, whose young still display reptilian claws on the tips of their wings.

The skeleton of a modern bird such as a pigeon still has much in common with that of an ancient dinosaur.

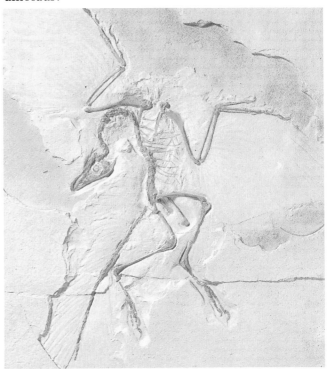

The fossil of archaeopteryx and how it looked with its feathers on!

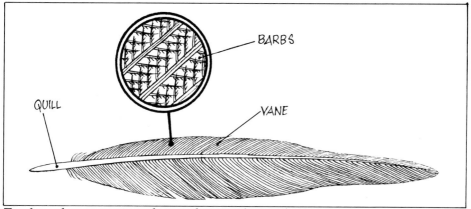

BARBS

QUILL

VANE

Feathers that are worn or damaged cannot be repaired so they are replaced when the bird moults. All birds moult completely once a year and some do it twice. Birds can fly at up to 276 km. per hour – and some (such as the Arctic tern) can travel 22,000 kilometres when migrating.

The feathers of a bird and its development of wings made true flight (not simply gliding which some reptiles and mammals can do) really possible. Some birds have immense wing-spans, particularly the Wandering Albatros of the Southern Ocean whose measurement from wing-tip to wing-tip can be over three metres.

Feathers also act to keep a bird warm and dry. To make sure there is always some waterproofing oil in its plumage, a bird regularly preens itself, distributing an oily secretion which it takes from a preen gland at its rear end.

Feather colour and design is also important for camouflage and for recognition by others of the same species. A snowy owl blends with an Arctic background and a Night hawk can look like a broken branch. Some unaggressive species of birds mimic stronger, tougher species by imitating their appearance.

Birds eat all manner of things and most species tend to specialize. Some are vegetarians, eating fruits, seeds, buds and leaves – humming-birds take nectar, geese graze on grass and grouse adore buds. Some are carnivores: a Secretary bird hunts

thrive on the flesh of dead animals, ospreys catch fish and the cuckoo is particularly fond of hairy caterpillars. Some birds are omnivorous and eat both vegetarian and meat dishes – the thrush likes fruit as well as slugs and the herring gull will swallow anything that looks the least bit edible on rubbish-tips often far inland.

The stomach of a bird is usually divided into two parts – the first is a sac called the crop and the second is a snakes, storks relish frogs, vultures

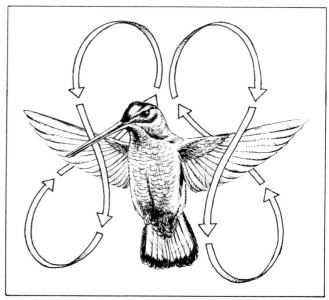

The humming bird's flight is in the pattern of a figure of 8 making it a living helicopter.

muscular food-grinding part called the gizzard. Stones and grit are swallowed by the bird and stored in the gizzard to help the grinding. The gizzard is another bit of anatomy handed down to birds from their reptile ancestors. Crocodiles also have gizzards.

Birds breathe not just by using lungs but also with the help of airsacs, thin-walled bags in the chest and abdomen. Parrots are the best talkers among birds but rarely have a vocabulary of more than 20 words. They do *not* understand what they are saying. Birdsong is principally an announcement that the singer is in possession of territory and serves to warn off intruders. It is an aid too in attracting mates. Perhaps some birds also sing just out of happiness.

Birds have excellent sight. They can see all colours except blue and violet. Like reptiles, birds have no external ear flaps but their sense of hearing is acute. Some birds such as owls and magpies can smell, but most probably cannot or only very poorly. Vultures are attracted to a dead animal by sight not smell.

Andrea holds a new-born chick in her hand.

The bird is a great voyager. It can find its way across the skies over vast distances with great accuracy. How does it do it? We know now that it possesses a small magnet in its skull which acts as a compass and that it has an instinctive knowledge of star positions. By releasing birds in a planetarium where star maps can be changed at will, scientists have shown that birds navigate by taking visual sightings of the stars and planets. During the day they plot the position of the sun, even when it is behind clouds. The human eye is not sensitive to polarized light.

Birds possess inbuilt biological clocks within their body and may also use a structure in their eye called the pecten much in the same way that a sailor uses a sextant for measuring the altitude of the sun. So, with compass, maps, clock and sextant, a bird is as well-equipped as any human navigator to go travelling.

An inbuilt biological clock, map, sextant and compass make the bird as well-equipped as any human navigator.